Add: _____   Job No: _____   Date: _____

I0461882

Panel No:	Batt Size	Date	Voltage	Current	Location

Add: Job No: Date:

Panel No:	Batt Size	Date	Voltage	Current	Location

Add: Job No: Date:

Panel No:	Batt Size	Date	Voltage	Current	Location

Add: Job No: Date:

Panel No:	Batt Size	Date	Voltage	Current	Location

Add: Job No: Date:

Panel No:	Batt Size	Date	Voltage	Current	Location

Add: Job No: Date:

Panel No:	Batt Size	Date	Voltage	Current	Location

Add: Job No: Date:

Panel No:	Batt Size	Date	Voltage	Current	Location

Add: Job No: Date:

Panel No:	Batt Size	Date	Voltage	Current	Location

Add: Job No: Date:

Panel No:	Batt Size	Date	Voltage	Current	Location

Add: Job No: Date:

Panel No:	Batt Size	Date	Voltage	Current	Location

Add: Job No: Date:

Panel No:	Batt Size	Date	Voltage	Current	Location

Add: Job No: Date:

Panel No:	Batt Size	Date	Voltage	Current	Location

Add: Job No: Date:

Panel No:	Batt Size	Date	Voltage	Current	Location

Add: Job No: Date:

Panel No:	Batt Size	Date	Voltage	Current	Location		

Add: Job No: Date:

Panel No:	Batt Size	Date	Voltage	Current	Location

Add: Job No: Date:

Panel No:	Batt Size	Date	Voltage	Current	Location		

Add: Job No: Date:

Panel No:	Batt Size	Date	Voltage	Current	Location

Add: Job No: Date:

Panel No:	Batt Size	Date	Voltage	Current	Location

Add: Job No: Date:

Panel No:	Batt Size	Date	Voltage	Current	Location

Add: Job No: Date:

Panel No:	Batt Size	Date	Voltage	Current	Location

Add: Job No: Date:

Panel No:	Batt Size	Date	Voltage	Current	Location

Add: Job No: Date:

Panel No:	Batt Size	Date	Voltage	Current	Location

Add: Job No: Date:

Panel No:	Batt Size	Date	Voltage	Current	Location

Add: Job No: Date:

Panel No:	Batt Size	Date	Voltage	Current	Location		

Add: Job No: Date:

Panel No:	Batt Size	Date	Voltage	Current	Location

Add: Job No: Date:

Panel No:	Batt Size	Date	Voltage	Current	Location

Add: Job No: Date:

Panel No:	Batt Size	Date	Voltage	Current	Location

Add: Job No: Date:

Panel No:	Batt Size	Date	Voltage	Current	Location

Add: Job No: Date:

Panel No:	Batt Size	Date	Voltage	Current	Location

Add: Job No: Date:

Panel No:	Batt Size	Date	Voltage	Current	Location

Add: Job No: Date:

Panel No:	Batt Size	Date	Voltage	Current	Location

Add: Job No: Date:

Panel No:	Batt Size	Date	Voltage	Current	Location

Add: Job No: Date:

Panel No:	Batt Size	Date	Voltage	Current	Location

Add: Job No: Date:

Panel No:	Batt Size	Date	Voltage	Current	Location

Add: Job No: Date:

Panel No:	Batt Size	Date	Voltage	Current	Location

Add: Job No: Date:

Panel No:	Batt Size	Date	Voltage	Current	Location

Add: Job No: Date:

Panel No:	Batt Size	Date	Voltage	Current	Location

Add: Job No: Date:

Panel No:	Batt Size	Date	Voltage	Current	Location

Add: Job No: Date:

Panel No:	Batt Size	Date	Voltage	Current	Location

Add: Job No: Date:

Panel No:	Batt Size	Date	Voltage	Current	Location	

Add: Job No: Date:

Panel No:	Batt Size	Date	Voltage	Current	Location

Add: Job No: Date:

Panel No:	Batt Size	Date	Voltage	Current	Location

Add: Job No: Date:

Panel No:	Batt Size	Date	Voltage	Current	Location

Add: Job No: Date:

Panel No:	Batt Size	Date	Voltage	Current	Location	

Add: Job No: Date:

Panel No:	Batt Size	Date	Voltage	Current	Location	

Add: Job No: Date:

Panel No:	Batt Size	Date	Voltage	Current	Location

Add: Job No: Date:

Panel No:	Batt Size	Date	Voltage	Current	Location

Add: Job No: Date:

Panel No:	Batt Size	Date	Voltage	Current	Location

Add: Job No: Date:

Panel No:	Batt Size	Date	Voltage	Current	Location

Add: Job No: Date:

Panel No:	Batt Size	Date	Voltage	Current	Location

Panel No:	Batt Size	Date	Voltage	Current	Location

Add: Job No: Date:

Panel No:	Batt Size	Date	Voltage	Current	Location

Add: Job No: Date:

Panel No:	Batt Size	Date	Voltage	Current	Location

Add: Job No: Date:

Panel No:	Batt Size	Date	Voltage	Current	Location

Add: Job No: Date:

Panel No:	Batt Size	Date	Voltage	Current	Location

Add: Job No: Date:

Panel No:	Batt Size	Date	Voltage	Current	Location		

Add: Job No: Date:

Panel No:	Batt Size	Date	Voltage	Current	Location

Add: Job No: Date:

Panel No:	Batt Size	Date	Voltage	Current	Location

Add: Job No: Date:

Panel No:	Batt Size	Date	Voltage	Current	Location

Add: Job No: Date:

Panel No:	Batt Size	Date	Voltage	Current	Location

Add: Job No: Date:

Panel No:	Batt Size	Date	Voltage	Current	Location

Add: Job No: Date:

Panel No:	Batt Size	Date	Voltage	Current	Location

Add: Job No: Date:

Panel No:	Batt Size	Date	Voltage	Current	Location

Add: Job No: Date:

Panel No:	Batt Size	Date	Voltage	Current	Location

Add: Job No: Date:

Panel No:	Batt Size	Date	Voltage	Current	Location

Add: Job No: Date:

Panel No:	Batt Size	Date	Voltage	Current	Location

Add: Job No: Date:

Panel No:	Batt Size	Date	Voltage	Current	Location

Add: Job No: Date:

Panel No:	Batt Size	Date	Voltage	Current	Location

Add: Job No: Date:

Panel No:	Batt Size	Date	Voltage	Current	Location

Add: Job No: Date:

Panel No:	Batt Size	Date	Voltage	Current	Location

Add: Job No: Date:

Panel No:	Batt Size	Date	Voltage	Current	Location

Add: Job No: Date:

Panel No:	Batt Size	Date	Voltage	Current	Location

Add: Job No: Date:

Panel No:	Batt Size	Date	Voltage	Current	Location	

Add: Job No: Date:

Panel No:	Batt Size	Date	Voltage	Current	Location

Add: Job No: Date:

Panel No:	Batt Size	Date	Voltage	Current	Location

Add: Job No: Date:

Panel No:	Batt Size	Date	Voltage	Current	Location

Add: Job No: Date:

Panel No:	Batt Size	Date	Voltage	Current	Location

Add: Job No: Date:

Panel No:	Batt Size	Date	Voltage	Current	Location

Add: Job No: Date:

Panel No:	Batt Size	Date	Voltage	Current	Location

Add: Job No: Date:

Panel No:	Batt Size	Date	Voltage	Current	Location

Add: Job No: Date:

Panel No:	Batt Size	Date	Voltage	Current	Location

Add: Job No: Date:

Panel No:	Batt Size	Date	Voltage	Current	Location

Add: Job No: Date:

Panel No:	Batt Size	Date	Voltage	Current	Location

Add: Job No: Date:

Panel No:	Batt Size	Date	Voltage	Current	Location

Add: Job No: Date:

Panel No:	Batt Size	Date	Voltage	Current	Location

Add: Job No: Date:

Panel No:	Batt Size	Date	Voltage	Current	Location

Add: Job No: Date:

Panel No:	Batt Size	Date	Voltage	Current	Location

Add: Job No: Date:

Panel No:	Batt Size	Date	Voltage	Current	Location

Add: Job No: Date:

Panel No:	Batt Size	Date	Voltage	Current	Location	

Add: Job No: Date:

Panel No:	Batt Size	Date	Voltage	Current	Location

Add: Job No: Date:

Panel No:	Batt Size	Date	Voltage	Current	Location

Add: Job No: Date:

Panel No:	Batt Size	Date	Voltage	Current	Location

Add: Job No: Date:

Panel No:	Batt Size	Date	Voltage	Current	Location

Add: Job No: Date:

Panel No:	Batt Size	Date	Voltage	Current	Location

Add: Job No: Date:

Panel No:	Batt Size	Date	Voltage	Current	Location

Add: Job No: Date:

Panel No:	Batt Size	Date	Voltage	Current	Location	

Add: Job No: Date:

Panel No:	Batt Size	Date	Voltage	Current	Location

Add: Job No: Date:

Panel No:	Batt Size	Date	Voltage	Current	Location

Add: Job No: Date:

Panel No:	Batt Size	Date	Voltage	Current	Location

Add: Job No: Date:

Panel No:	Batt Size	Date	Voltage	Current	Location

Add: Job No: Date:

Panel No:	Batt Size	Date	Voltage	Current	Location

Add: Job No: Date:

Panel No:	Batt Size	Date	Voltage	Current	Location

Add:

Add: Job No: Date:

Panel No:	Batt Size	Date	Voltage	Current	Location

Add: Job No: Date:

Add: Job No: Date:

Panel No:	Batt Size	Date	Voltage	Current	Location

Add: Job No: Date:

Add: Job No: Date:

Panel No:	Batt Size	Date	Voltage	Current	Location

Add: Job No: Date:

Panel No:	Batt Size	Date	Voltage	Current	Location

Add: Job No: Date:

Add: Job No: Date:

Panel No:	Batt Size	Date	Voltage	Current	Location

Add: Job No: Date:

Panel No:	Batt Size	Date	Voltage	Current	Location

Add:

Add: Job No: Date:

Panel No:	Batt Size	Date	Voltage	Current	Location

Add: Job No: Date:

Panel No:	Batt Size	Date	Voltage	Current	Location

Add:

Add: Job No: Date:

Panel No:	Batt Size	Date	Voltage	Current	Location

Add: Job No: Date:

Panel No:	Batt Size	Date	Voltage	Current	Location

Add: Job No: Date:

Panel No:	Batt Size	Date	Voltage	Current	Location

Add: Job No: Date:

Panel No:	Batt Size	Date	Voltage	Current	Location

Add: Job No: Date:

Panel No:	Batt Size	Date	Voltage	Current	Location

Add: Job No: Date:

Panel No:	Batt Size	Date	Voltage	Current	Location

Add:

Add: Job No: Date:

Panel No:	Batt Size	Date	Voltage	Current	Location		

Add: Job No: Date:

Panel No:	Batt Size	Date	Voltage	Current	Location		

Add: Job No: Date:

Panel No:	Batt Size	Date	Voltage	Current	Location

Add: Job No: Date:

Panel No:	Batt Size	Date	Voltage	Current	Location

Add: Job No: Date:

Panel No:	Batt Size	Date	Voltage	Current	Location

Add: Job No: Date:

Panel No:	Batt Size	Date	Voltage	Current	Location

Add: Job No: Date:

Panel No:	Batt Size	Date	Voltage	Current	Location

Add: Job No: Date:

Panel No:	Batt Size	Date	Voltage	Current	Location

Add: Job No: Date:

Panel No:	Batt Size	Date	Voltage	Current	Location

Add: Job No: Date:

Panel No:	Batt Size	Date	Voltage	Current	Location

Add:

Add: Job No: Date:

Panel No:	Batt Size	Date	Voltage	Current	Location

Add: Job No: Date:

Panel No:	Batt Size	Date	Voltage	Current	Location

Add: Job No: Date:

Panel No:	Batt Size	Date	Voltage	Current	Location

Add: Job No: Date:

Panel No:	Batt Size	Date	Voltage	Current	Location

Add: Job No: Date:

Add: Job No: Date:

Panel No:	Batt Size	Date	Voltage	Current	Location

Add: Job No: Date:

Panel No:	Batt Size	Date	Voltage	Current	Location

Add:

Add: Job No: Date:

Panel No:	Batt Size	Date	Voltage	Current	Location

Add: Job No: Date:

Panel No:	Batt Size	Date	Voltage	Current	Location

Add: Job No: Date:

Panel No:	Batt Size	Date	Voltage	Current	Location			

Add: Job No: Date:

Panel No:	Batt Size	Date	Voltage	Current	Location

Add:

Add: Job No: Date:

Panel No:	Batt Size	Date	Voltage	Current	Location

Add: Job No: Date:

Panel No:	Batt Size	Date	Voltage	Current	Location

Add:

Add: Job No: Date:

Panel No:	Batt Size	Date	Voltage	Current	Location

Add: Job No: Date:

Panel No:	Batt Size	Date	Voltage	Current	Location

Add: Job No: Date:

Panel No:	Batt Size	Date	Voltage	Current	Location

Add: Job No: Date:

Panel No:	Batt Size	Date	Voltage	Current	Location

Add: Job No: Date:

Panel No:	Batt Size	Date	Voltage	Current	Location	

Add: Job No: Date:

Panel No:	Batt Size	Date	Voltage	Current	Location

Add: Job No: Date:

Panel No:	Batt Size	Date	Voltage	Current	Location

Add: Job No: Date:

Panel No:	Batt Size	Date	Voltage	Current	Location

Add: Job No: Date:

Add: Job No: Date:

Panel No:	Batt Size	Date	Voltage	Current	Location

Add: Job No: Date:

Panel No:	Batt Size	Date	Voltage	Current	Location	

Add: Job No: Date:

Panel No:	Batt Size	Date	Voltage	Current	Location

Add: Job No: Date:

Panel No:	Batt Size	Date	Voltage	Current	Location

Add: Job No: Date:

Panel No:	Batt Size	Date	Voltage	Current	Location

Add: Job No: Date:

Panel No:	Batt Size	Date	Voltage	Current	Location

Add: Job No: Date:

Panel No:	Batt Size	Date	Voltage	Current	Location

Add: Job No: Date:

Panel No:	Batt Size	Date	Voltage	Current	Location

Add: Job No: Date:

Panel No:	Batt Size	Date	Voltage	Current	Location

Add: Job No: Date:

Panel No:	Batt Size	Date	Voltage	Current	Location

Add: Job No: Date:

Panel No:	Batt Size	Date	Voltage	Current	Location

Add: Job No: Date:

Panel No:	Batt Size	Date	Voltage	Current	Location

Add: Job No: Date:

Panel No:	Batt Size	Date	Voltage	Current	Location

Add: Job No: Date:

Panel No:	Batt Size	Date	Voltage	Current	Location

Add: Job No: Date:

Panel No:	Batt Size	Date	Voltage	Current	Location

Add: Job No:

Add: Job No: Date:

Panel No:	Batt Size	Date	Voltage	Current	Location	

Add:

Add: Job No: Date:

Panel No:	Batt Size	Date	Voltage	Current	Location

Add: Job No: Date:

Panel No:	Batt Size	Date	Voltage	Current	Location

Add: Job No: Date:

Panel No:	Batt Size	Date	Voltage	Current	Location

Add: Job No: Date:

Panel No:	Batt Size	Date	Voltage	Current	Location

Add: Job No: Date:

Panel No:	Batt Size	Date	Voltage	Current	Location

Add: Job No: Date:

Panel No:	Batt Size	Date	Voltage	Current	Location

Add:

Add: Job No: Date:

Panel No:	Batt Size	Date	Voltage	Current	Location

Add: Job No: Date:

Panel No:	Batt Size	Date	Voltage	Current	Location

Add: Job No: Date:

Panel No:	Batt Size	Date	Voltage	Current	Location

Add: Job No: Date:

Panel No:	Batt Size	Date	Voltage	Current	Location

Add: Job No: Date:

Panel No:	Batt Size	Date	Voltage	Current	Location

Add: Job No: Date:

Panel No:	Batt Size	Date	Voltage	Current	Location

Add:

Add: Job No: Date:

Panel No:	Batt Size	Date	Voltage	Current	Location

Add: Job No: Date:

Panel No:	Batt Size	Date	Voltage	Current	Location

Add: Job No: Date:

Panel No:	Batt Size	Date	Voltage	Current	Location

Add: Job No: Date:

Panel No:	Batt Size	Date	Voltage	Current	Location

Add: Job No: Date:

Panel No:	Batt Size	Date	Voltage	Current	Location

Add: Job No: Date:

Panel No:	Batt Size	Date	Voltage	Current	Location

Add:

Add: Job No: Date:

Panel No:	Batt Size	Date	Voltage	Current	Location

Add: Job No: Date:

Panel No:	Batt Size	Date	Voltage	Current	Location

Add: Job No: Date:

Panel No:	Batt Size	Date	Voltage	Current	Location

Add: Job No: Date:

Panel No:	Batt Size	Date	Voltage	Current	Location

Add: Job No: Date:

Panel No:	Batt Size	Date	Voltage	Current	Location

Add: Job No: Date:

Panel No:	Batt Size	Date	Voltage	Current	Location

Add: Job No: Date:

Panel No:	Batt Size	Date	Voltage	Current	Location

Add: Job No: Date:

Panel No:	Batt Size	Date	Voltage	Current	Location

Add: Job No: Date:

Panel No:	Batt Size	Date	Voltage	Current	Location

Add: Job No: Date:

Panel No:	Batt Size	Date	Voltage	Current	Location

Add: Job No: Date:

Panel No:	Batt Size	Date	Voltage	Current	Location

Add: Job No: Date:

Panel No:	Batt Size	Date	Voltage	Current	Location

Add: Job No: Date:

Panel No:	Batt Size	Date	Voltage	Current	Location

Add: Job No: Date:

Panel No:	Batt Size	Date	Voltage	Current	Location

Add:

Add: Job No: Date:

Panel No:	Batt Size	Date	Voltage	Current	Location

Add: Job No: Date:

Panel No:	Batt Size	Date	Voltage	Current	Location

Add: Job No: Date:

Panel No:	Batt Size	Date	Voltage	Current	Location

Add: Job No: Date:

Panel No:	Batt Size	Date	Voltage	Current	Location

Add:

Add: Job No: Date:

Panel No:	Batt Size	Date	Voltage	Current	Location			

Add: Job No: Date:

Panel No:	Batt Size	Date	Voltage	Current	Location